LITHIUM MINING IN AUSTRALIA AND CHINA

An insight into mining activities in both countries

By

VALATI CLIFFORD

TABLE OF CONTENTS

INTRODUCTION

- Overview of lithium mining industry in Australia and China
- Brief history of lithium and its uses
- Objectives of the book

CHAPTER ONE

- The Lithium Mining Industry in Australia
- Overview of the Australian lithium mining industry
- Major lithium mines and their production
- Economic impacts of lithium mining in Australia
- Environmental and social concerns surrounding lithium mining in Australia
- Future outlook for the Australian lithium mining industry

CHAPTER TWO

- The Lithium Mining Industry in China

- Overview of the Chinese lithium mining industry
- Major lithium mines and their production
- Economic impacts of lithium mining in China

CHAPTER THREE

- Environmental and social concerns surrounding lithium mining in China
- Future outlook for the Chinese lithium mining industry
- Comparison of Lithium Mining in Australia and China
- Production volumes and trends in Australia and China
- Cost structures and competitiveness of lithium mining in Australia and China

CHAPTER FOUR

- Environmental and social impacts of lithium mining in Australia and China
- Regulatory frameworks for lithium mining in Australia and China

CHAPTER FIVE
- Technological developments and innovations in lithium mining in Australia and China
- The Future of Lithium Mining in Australia and China

CHAPTER SIX
- Emerging trends and challenges in the lithium mining industry
- Opportunities for growth and investment in the Australian and Chinese lithium mining industries
- Alternative sources of lithium and their potential impacts on the industry
- Prospects for sustainable and responsible lithium mining in Australia and China

CONCLUSION
- Summary of key findings
- Implications for policy makers, investors, and industry stakeholders
- Future prospects for the lithium mining industry in Australia and China

INTRODUCTION
AUSTRALIA

Australia is the world's second-largest producer of lithium, accounting for approximately 48% of global production in 2020. The country has several large lithium deposits, with the largest being the Greenbushes lithium mine in Western Australia, which is operated by Talison Lithium. The mine produces approximately 40% of the world's lithium, and there are several other lithium projects underway in Australia, including the Mount Marion mine and the Mt Cattlin mine.

The Australian lithium mining industry has seen significant growth in recent years, driven by the increasing demand for lithium-ion batteries used in electric vehicles and renewable energy storage systems. The industry has become an important contributor to the Australian economy, providing employment opportunities and driving economic growth in regional areas. However, the industry has also faced criticism from environmental groups over the potential impacts of mining on local ecosystems and communities.

CHINA

China is the world's largest producer of lithium, accounting for approximately 62% of global production in 2020. The country has several large lithium deposits, with the largest being the Zhabuye Salt Lake in Tibet, which is operated by Tibet Mineral Development Co. Ltd.

The Chinese lithium mining industry has grown rapidly in recent years, driven by the increasing demand for lithium-ion batteries used in electric vehicles and consumer electronics. The industry has become a key player in the global lithium market, and Chinese companies have invested heavily in lithium projects overseas, particularly in South America and Africa.

The Chinese government has supported the growth of the lithium mining industry through policies and incentives aimed at promoting the development of the electric vehicle industry.

However, the industry has also faced criticism over environmental concerns, particularly in the extraction of lithium from salt flats, which can have significant impacts on local ecosystems and water resources.

Overall, both Australia and China are major players in the global lithium mining industry, with significant deposits and growing production. However, the industry faces challenges related to environmental and social impacts, and the development of sustainable and responsible mining practices will be critical for its long-term success.

THE RELATIONSHIP BETWEEN AUSTRALIA LITHIUM MINING INDUSTRY AND CHINA

The relationship between the Australian lithium mining industry and China is complex and multifaceted, driven by economic, political, and strategic factors.

China is the world's largest consumer of lithium, driven by the rapid growth of its electric vehicle industry and the increasing demand for renewable energy storage systems. As a result, China has become a major buyer of lithium from Australia, which is one of the world's largest producers of lithium. In 2020, China accounted for approximately 70% of Australia's lithium exports, valued at around AUD 1.1 billion.

The relationship between the two countries has been strengthened by the signing of several agreements and partnerships aimed at promoting cooperation in the lithium mining industry. For example, in 2019, Australia's Pilbara Minerals signed a binding agreement with China's Jiangxi Ganfeng Lithium Co. Ltd. to develop the Pilgangoora lithium-tantalum project in Western Australia. Under the agreement, Ganfeng Lithium acquired a 43.1% stake in Pilbara Minerals and agreed to purchase a fixed quantity of lithium concentrate each year.

However, the relationship between Australia and China has also been strained in recent years due to political and strategic tensions, particularly in the wake of the COVID-19 pandemic and increasing concerns over national security. This has led to disruptions in trade and investment flows between the two countries, including in the lithium mining industry. For example, in 2020, China threatened to impose tariffs on Australian lithium exports in retaliation for Australia's call for an investigation into the origins of the COVID-19 pandemic.

Overall, the relationship between the Australian lithium mining industry and China is characterized by a complex mix of economic, political, and strategic factors, which will continue to shape the future of the industry in both countries.

BRIEF HISTORY OF LITHIUM AND ITS USES

Lithium is a chemical element with the symbol Li and atomic number 3. It is a soft, silvery-white metal that belongs to the alkali metal group of elements. Lithium was first discovered in 1817 by the Swedish chemist Johan August Arfwedson, who identified the element in a sample of petalite ore.

Lithium has a wide range of uses, including in the production of ceramics, glass, and lubricants. However, its most important use is in the manufacture of rechargeable batteries, particularly for electric vehicles and portable electronic devices.

The use of lithium in batteries dates back to the 1970s, when researchers at Exxon developed the first lithium-ion battery. However, it was not until the 1990s that lithium-ion batteries became commercially viable, and they have since become the dominant technology in the global battery market.

In addition to its use in batteries, lithium has also been used in the treatment of bipolar disorder, a mental illness characterized by alternating periods of depression and mania. Lithium was first used for this purpose in the 1940s, and it remains an important medication for the treatment of bipolar disorder to this day.

Other uses of lithium include in the production of alloys, such as aluminum-lithium alloys used in the aerospace industry, and in the production of synthetic rubber and polymers. Lithium also has applications in nuclear physics, where it is used as a coolant in nuclear reactors.

Overall, the history of lithium is one of discovery and innovation, driven by the element's unique properties and its wide range of uses. From its discovery in the 19th century to its current role as a critical component of the global battery market, lithium has played an important role in shaping the modern world.

OBJECTIVES OF THE BOOK

The book on the topic of lithium mining activities in both Australia and China has several objectives, including:

Providing an overview of the lithium mining industry in Australia and China: The book aims to provide readers with a comprehensive understanding of the current state of the lithium mining industry in both countries, including the size and scope of the industry, the key players, and the major trends and challenges.

Examining the economic and political factors driving lithium mining in both countries: The book aims to explore the economic and political factors that are driving the growth of the lithium mining industry in Australia and China, including the increasing demand for lithium due to the rise of electric vehicles and renewable energy systems, as well as the strategic implications of this demand.

Analyzing the environmental and social impacts of lithium mining: The book aims to examine the environmental and social impacts of lithium mining activities in both Australia and China, including the potential risks and challenges associated with lithium extraction, such as water use, waste disposal, and land use.

Evaluating the technological innovations and advancements in lithium mining: The book aims to assess the technological innovations and advancements in lithium mining activities, including the development of new extraction methods, processing techniques, and recycling technologies.

Identifying the future prospects and challenges of lithium mining in both countries: The book aims to provide readers with insights into the future prospects and challenges of lithium mining activities in both Australia and China, including the potential for growth, emerging risks, and opportunities for collaboration and cooperation between the two countries.

Overall, the book aims to provide a comprehensive and balanced analysis of the lithium mining industry in both Australia and China, highlighting the opportunities and challenges associated with this critical sector of the global economy.

CHAPTER ONE

THE LITHIUM MINING INDUSTRY IN AUSTRALIA

The Australian lithium mining industry has experienced significant growth over the past decade, driven by the increasing demand for lithium-ion batteries used in electric vehicles, smartphones, and other electronic devices. Australia is now the largest producer of lithium globally, accounting for around 52% of the world's total production in 2020.

Lithium is primarily extracted from hard-rock spodumene deposits found in Western Australia, particularly the Greenbushes Lithium mine, which is the world's largest lithium mine. Other major lithium mines in Australia include Mt. Cattlin, Mt. Marion, and Bald Hill.

In addition to hard-rock mining, Australia also has significant lithium brine deposits, particularly in the Northern Territory and Western Australia. Lithium brine extraction involves pumping salty water from underground reservoirs and evaporating the water to concentrate the lithium.

The Australian lithium mining industry has faced some challenges in recent years, including a slump in lithium prices and the COVID-19 pandemic. However, the industry has shown resilience and is expected to continue growing as demand for electric vehicles and renewable energy storage systems increases.

Australia is also investing in developing downstream processing capabilities, such as lithium hydroxide and lithium carbonate production, which are higher-value products used in battery manufacturing. This is expected to increase the value of Australia's lithium exports and create additional job opportunities in the industry.

The Australian government has identified lithium as a key strategic mineral and is actively promoting investment in the sector through initiatives such as the Critical Minerals Strategy and the Modern Manufacturing Initiative. This support, combined with Australia's abundant lithium resources and well-established mining industry, positions Australia as a significant player in the global lithium market.

MAJOR LITHIUM MINES AND THEIR PRODUCTION

Australia is the largest producer of lithium in the world, and the majority of its lithium production comes from hard-rock spodumene deposits found in Western Australia. Here are the major lithium mines in Australia and their production:

Greenbushes Lithium Mine:
Located in Western Australia, the Greenbushes Lithium Mine is the world's largest lithium mine. It is owned and operated by Talison Lithium, a subsidiary of Chinese company Tianqi Lithium. The mine produces both lithium concentrate and lithium hydroxide. In 2020, the mine produced approximately 62,000 tonnes of lithium

concentrate and 24,000 tonnes of lithium hydroxide.

Mt. Cattlin Lithium Mine:
Mt. Cattlin is another major lithium mine in Western Australia, owned and operated by Galaxy Resources. The mine produces lithium concentrate, which is exported to China for processing. In 2020, the mine produced approximately 35,000 tonnes of lithium concentrate.

Mt. Marion Lithium Mine:
Located in Western Australia, the Mt. Marion Lithium Mine is a joint venture between Chinese company Ganfeng Lithium, Australian miner Mineral Resources, and Neometals. The mine produces lithium concentrate, which is exported to China for processing. In 2020, the mine produced approximately 38,000 tonnes of lithium concentrate.

Bald Hill Lithium Mine:

The Bald Hill Lithium Mine is located in Western Australia and is a joint venture between Australian miner Tawana Resources and Singapore-based Alliance Mineral Assets. The mine produces lithium concentrate, which is exported to China for processing. In 2020, the mine produced approximately 35,000 tonnes of lithium concentrate.

Pilgangoora Lithium Mine:
The Pilgangoora Lithium Mine is located in Western Australia and is owned by Pilbara Minerals. The mine produces spodumene concentrate, which is processed into lithium hydroxide and lithium carbonate. In 2020, the mine produced approximately 96,000 tonnes of spodumene concentrate.

Wodgina Lithium Mine:
Located in Western Australia, the Wodgina Lithium Mine is owned by Mineral Resources. The mine produces spodumene concentrate, which is exported to China for processing. In

2020, the mine produced approximately 80,000 tonnes of spodumene concentrate.

Earl Grey Lithium Mine:
The Earl Grey Lithium Mine is a joint venture between Kidman Resources and Chilean company SQM. The mine is located in Western Australia and produces spodumene concentrate, which is exported to China for processing. In 2020, the mine produced approximately 38,000 tonnes of spodumene concentrate.

Overall, the Australian lithium mining industry is highly concentrated, with a few major players dominating the market. The majority of the lithium produced is exported to China for processing, highlighting the importance of the global supply chain in the lithium industry. However, there is growing interest in developing downstream processing capabilities in Australia, which could increase the value of Australia's lithium exports and create additional job opportunities in the industry.

ECONOMIC IMPACTS OF LITHIUM MINING IN AUSTRALIA

The rise of the electric vehicle (EV) industry has created a significant demand for lithium, a key component of lithium-ion batteries. Australia is the largest producer of lithium globally, and the economic impacts of lithium mining in Australia have been significant. In this article, we will explore the economic impacts of lithium mining in Australia, including the benefits and challenges.

BENEFITS OF LITHIUM MINING IN AUSTRALIA

1.Job Creation: Lithium mining has created significant job opportunities in Australia, particularly in Western Australia where the majority of lithium mining occurs. According to the Australian Bureau of Statistics, the mining industry employed approximately 246,100 people in 2020, with the majority of these jobs being in Western Australia. Lithium mining has also created jobs in related industries such as transport, logistics, and engineering.

2.Increased Export Revenue: Lithium mining has contributed significantly to Australia's export revenue. According to the Australian Government, lithium was Australia's fourth-largest mineral export in 2020, with a value of $1.8 billion. The demand for lithium is expected to continue growing, which could lead to increased export revenue in the future.

3.Investment and Innovation: The lithium mining industry has attracted significant investment and innovation in Australia. Major mining companies such as BHP, Rio Tinto, and Fortescue Metals Group have invested in lithium mining projects in Australia, and there has also been investment in downstream processing capabilities. This investment and innovation have the potential to create new technologies and jobs in Australia.

CHALLENGES OF LITHIUM MINING IN AUSTRALIA

1.Environmental Impact: Lithium mining can have significant environmental impacts, including land degradation, water pollution, and carbon emissions. In particular, the process of extracting lithium from hard-rock spodumene deposits can involve large amounts of water and energy. Lithium brine extraction can also have significant impacts on groundwater resources and ecosystems.

2.Community Impacts: Lithium mining can have impacts on local communities, particularly indigenous communities. Mining can impact cultural heritage sites, and there have been concerns about the impact of mining on water resources and the health of communities.

3.Volatility of the Market: The lithium market can be volatile, with prices fluctuating depending on supply and demand. This can create uncertainty for companies and communities involved in the lithium mining industry.

CONCLUSION

The economic impacts of lithium mining in Australia have been significant, with job creation, increased export revenue, and investment and innovation. However, there are also challenges associated with lithium mining, including environmental impacts and community concerns. It is essential for the industry to address these challenges and work with communities and stakeholders to ensure sustainable and responsible mining practices. The Australian Government has identified lithium as a key strategic mineral and is actively promoting investment in the sector, while also supporting initiatives such as the Critical Minerals Strategy and the Modern Manufacturing Initiative. The lithium mining industry has the potential to create economic benefits for Australia, while also contributing to the global transition towards sustainable and renewable energy sources.

ENVIRONMENTAL AND SOCIAL CONCERNS SURROUNDING LITHIUM MINING IN AUSTRALIA

The growth of the electric vehicle industry has created a surge in demand for lithium, a key component of lithium-ion batteries. Australia is one of the largest producers of lithium globally, and while the economic benefits of lithium mining are evident, there are also environmental and social concerns surrounding the industry. In this article, we will explore the environmental and social concerns surrounding lithium mining in Australia.

ENVIRONMENTAL CONCERNS

1.Land Degradation: Lithium mining can lead to significant land degradation, particularly in areas where hard-rock spodumene deposits are mined. The mining process involves removing large amounts of rock and soil, which can result in soil erosion, loss of vegetation, and destruction of habitats for wildlife.

2.Water Usage: The process of extracting lithium from hard-rock spodumene deposits can involve large amounts of water. This can lead to water scarcity in regions that are already experiencing water stress, such as Western Australia.

3.Carbon Emissions: Lithium mining can also result in significant carbon emissions. The mining process involves the use of heavy machinery and equipment, which requires fossil fuels to operate. Additionally, the transportation of lithium products to markets around the world also contributes to carbon emissions.

SOCIAL CONCERNS

Impact on Indigenous Communities: Lithium mining can impact indigenous communities, particularly in Australia, where many mining sites are located on traditional lands. There have been concerns about the impact of mining on cultural heritage sites, as well as the impact on the health and wellbeing of indigenous communities.

1.Local Communities: Lithium mining can also impact local communities, particularly in terms of noise and dust pollution, as well as increased traffic and pressure on local infrastructure. In some cases, mining companies have also been accused of displacing local communities.

2.Labor Practices: The lithium mining industry has also faced scrutiny over labor practices, particularly in relation to the use of temporary and casual labor, which can result in job insecurity and low wages.

CONCLUSION

The environmental and social concerns surrounding lithium mining in Australia are significant and cannot be ignored. The industry must take steps to address these concerns and ensure that mining practices are sustainable and responsible. This includes implementing measures to minimize land degradation, reducing water usage, and transitioning to low-carbon energy sources. It also requires engaging with indigenous communities and local communities to ensure that their rights and concerns are respected. Furthermore, there must be a focus on ensuring fair and safe working conditions for all workers in the industry. By addressing these concerns, the lithium mining industry in Australia can contribute to a sustainable and equitable future for all.

FUTURE OUTLOOK FOR THE AUSTRALIAN LITHIUM MINING INDUSTRY

The Australian lithium mining industry has seen significant growth in recent years, driven by the increasing demand for lithium-ion batteries for electric vehicles and renewable energy storage. As a result, the future outlook for the Australian lithium mining industry is positive, with significant potential for growth and innovation.

In this article, we will explore the future outlook for the Australian lithium mining industry.

1.Increasing Demand: The demand for lithium is expected to continue growing in the coming years, driven by the increasing adoption of electric vehicles and renewable energy storage systems. This is expected to create significant opportunities for the Australian lithium mining industry, as Australia is one of the largest producers of lithium globally.

2.Expansion of Lithium Processing: There is significant potential for the expansion of downstream processing capabilities in Australia, which could create new jobs and investment opportunities. This includes the development of lithium refining facilities, which would allow for the production of higher-value lithium products.

3.Focus on Sustainable Practices: There is increasing pressure on the mining industry to adopt sustainable and responsible practices, and the lithium mining industry is no exception. This includes implementing measures to minimize environmental impacts, engaging with local communities, and ensuring fair and safe working conditions for workers in the industry.

4.Investment in Technology and Innovation: The Australian lithium mining industry is also poised for significant investment in technology and innovation. This includes the development of new extraction methods, such as direct lithium extraction, which could significantly reduce water usage and environmental impacts.

5.Government Support: The Australian Government has identified lithium as a key strategic mineral and is actively promoting investment in the sector. This includes initiatives such as the Critical Minerals Strategy and the Modern Manufacturing Initiative, which provide support for research and development, infrastructure, and investment in the lithium mining industry.

CONCLUSION

The future outlook for the Australian lithium mining industry is positive, with significant potential for growth and innovation. The increasing demand for lithium, the expansion of downstream processing capabilities, a focus on sustainable practices, investment in technology and innovation, and government support all indicate a bright future for the industry. However, it is crucial that the industry continues to address environmental and social concerns, engage with local communities, and ensure fair and safe working conditions for all workers. By doing so, the Australian lithium mining industry can play a significant role in the global transition towards sustainable and renewable energy sources.

CHAPTER TWO

THE LITHIUM MINING INDUSTRY IN CHINA

China is one of the world's largest producers of lithium, a key component in lithium-ion batteries used in electric vehicles and other energy storage systems. The country's lithium mining industry has seen significant growth in recent years, driven by the increasing demand for lithium-ion batteries. In this article, we will explore the overviews of the Chinese lithium mining industry.

Production and Reserves: China has significant lithium reserves, with an estimated 3.2 million metric tons of lithium resources, accounting for approximately 11% of the world's total lithium resources. China is also the world's largest producer of lithium, accounting for approximately 60% of global production in 2020.

1.Lithium Production Methods: The majority of lithium production in China comes from brine deposits, particularly in the Qinghai-Tibet Plateau. The extraction process involves pumping brine from underground aquifers and allowing the water to evaporate, leaving behind lithium salts. Hard-rock lithium mining also takes place in China, with the largest hard-rock lithium mine in the country located in Yichun, Jiangxi Province.

2.Major Lithium Producers: China's major lithium producers include Tianqi Lithium, Ganfeng Lithium, and Qinghai Salt Lake Industry Group. These companies have significant operations in China and also have invested in lithium mining and processing operations globally.

3.Government Support: The Chinese government has identified lithium as a strategic mineral and is actively promoting investment in the industry. This includes policies such as the New Energy Vehicle Industry Development

Plan, which aims to support the development of electric vehicles and the lithium-ion battery industry.

4.Environmental and Social Concerns: The Chinese lithium mining industry has faced criticism for its environmental and social impacts. The extraction of lithium from brine deposits can lead to significant water usage, which can exacerbate water scarcity in regions that are already experiencing water stress. There are also concerns about the impact of mining on local ecosystems and the displacement of local communities.

CONCLUSION
China's lithium mining industry has seen significant growth in recent years, driven by the increasing demand for lithium-ion batteries. The country's significant reserves and production capabilities make it a major player in the global lithium mining industry. However, the industry must also address environmental and social concerns, including the impact on local

ecosystems and communities and the water usage associated with lithium extraction. By doing so, the Chinese lithium mining industry can contribute to a sustainable and responsible global transition towards renewable energy sources.

MAJOR LITHIUM MINES AND THEIR PRODUCTION

Lithium is a highly valued mineral due to its use in lithium-ion batteries, which power electric vehicles and other energy storage systems. The global demand for lithium has increased significantly in recent years, driving the growth of the lithium mining industry. In this article, we will explore some of the major lithium mines and their production around the world.

GREENBUSHES LITHIUM MINE, AUSTRALIA

The Greenbushes lithium mine, located in Western Australia, is one of the world's largest hard-rock lithium mines. The mine is owned by

Talison Lithium, a subsidiary of China's Tianqi Lithium, and has been in operation since 1983. In 2020, the mine produced approximately 77,000 metric tons of lithium concentrate.

SALAR DE ATACAMA LITHIUM MINE, CHILE

The Salar de Atacama lithium mine, located in the Atacama Desert in northern Chile, is one of the world's largest lithium brine mines. The mine is operated by SQM (Sociedad Química y Minera de Chile), a Chilean chemical company. In 2020, the mine produced approximately 71,000 metric tons of lithium carbonate equivalent.

ALBEMARLE SILVER PEAK LITHIUM MINE, USA

The Silver Peak lithium mine, located in Nevada, USA, is one of the oldest lithium mines in the world. The mine is owned by Albemarle, a global specialty chemicals company. The Silver Peak mine produces lithium using a unique evaporation process and has a capacity of approximately 5,000 metric tons per year.

LITHIUM AMERICAS CAUCHARI-OLAROZ LITHIUM PROJECT, ARGENTINA

The Cauchari-Olaroz lithium project, located in Argentina's Jujuy province, is a joint venture between Lithium Americas and China's Ganfeng Lithium. The project is still under development, with the first phase expected to start production in 2022. The project is expected to have an annual production capacity of approximately 40,000 metric tons of lithium carbonate.

JADAR LITHIUM PROJECT, SERBIA

The Jadar lithium project, located in Serbia, is owned by Rio Tinto and is currently in the development phase. The project is expected to produce lithium using a unique process that extracts lithium from jadarite, a rare mineral found only in Serbia. The project is expected to have a production capacity of approximately 55,000 metric tons of lithium carbonate per year.

CONCLUSION

The global demand for lithium is expected to continue growing in the coming years, driven by the increasing adoption of electric vehicles and renewable energy storage systems. The major lithium mines around the world, including the Greenbushes lithium mine in Australia, the Salar de Atacama lithium mine in Chile, the Albemarle Silver Peak lithium mine in the USA, the Lithium Americas Cauchari-Olaroz lithium project in Argentina, and the Jadar lithium project in Serbia, are critical to meeting this demand. The continued development and investment in these mines, as well as the exploration of new lithium deposits, will be crucial in meeting the world's growing demand for lithium and supporting the transition towards sustainable and renewable energy sources.

ECONOMIC IMPACTS OF LITHIUM MINING IN CHINA

China is the world's largest producer of lithium, accounting for approximately two-thirds of the global production. The country's lithium mining industry has significant economic impacts, not only on China but also on the global market for lithium.

ECONOMIC BENEFITS OF LITHIUM MINING IN CHINA

1.Job Creation: The lithium mining industry in China provides employment opportunities for thousands of people, both directly and indirectly. This includes workers involved in mining, processing, transportation, and related industries.

2.Revenue Generation: The Chinese government collects revenue from the lithium mining industry through taxes, royalties, and other fees. This revenue can be used to fund social programs and infrastructure development.

3.Export Earnings: China exports a significant portion of the lithium it produces to other countries, earning foreign exchange revenue. This helps to support the country's balance of payments and strengthen its economy.

4.Industrial Development: The lithium mining industry in China supports the growth of related industries, such as battery manufacturing and electric vehicle production. This promotes the development of new technologies and supports the country's transition towards clean energy.

5.Innovation: The lithium mining industry in China is driving innovation in mining and processing technologies. This is leading to more efficient and sustainable production methods, which can be applied to other mineral resources as well.

CHALLENGES OF LITHIUM MINING IN CHINA

Environmental Concerns: The lithium mining industry can have significant environmental impacts, including water pollution, soil degradation, and greenhouse gas emissions. The Chinese government has implemented regulations and policies to mitigate these impacts, but there is still room for improvement.

1.Social Impacts: Lithium mining can also have social impacts, such as the displacement of communities and the disruption of traditional livelihoods. The Chinese government and mining companies must work to minimize these impacts and ensure that affected communities are properly compensated.

2.Market Volatility: The lithium market is highly volatile, with prices fluctuating significantly depending on global supply and demand. This can have an impact on the profitability of lithium mining companies and the revenue generated by the Chinese government.

3.Resource Depletion: Lithium is a finite resource, and the rapid growth of the lithium mining industry could lead to resource depletion in the future. This highlights the need for sustainable mining practices and the development of alternative sources of energy storage.

CONCLUSION

The economic impacts of lithium mining in China are significant, providing employment opportunities, revenue generation, and industrial development. However, the industry also presents challenges, including environmental and social impacts, market volatility, and resource depletion. The Chinese government and mining companies must work together to address these challenges and promote sustainable mining practices. In doing so, they can support the country's economic development while also contributing to the global transition towards clean energy.

ENVIRONMENTAL AND SOCIAL CONCERNS SURROUNDING LITHIUM MINING IN CHINA

As the world's largest producer of lithium, China's lithium mining industry has significant environmental and social impacts. Here are some of the main concerns surrounding lithium mining in China:

ENVIRONMENTAL CONCERNS

1.Water Usage: Lithium mining requires significant amounts of water, which can put pressure on already scarce water resources in some regions. This can lead to water shortages and negatively impact local communities and ecosystems.

2.Land Use: Lithium mining often involves large-scale excavation and land disturbance, which can lead to soil erosion, habitat loss, and other environmental impacts.

3.Air Pollution: The processing of lithium ores can release pollutants into the air, such as dust, sulfur dioxide, and carbon dioxide, which can contribute to air pollution and climate change.

4.Chemical Contamination: The processing of lithium ores involves the use of chemicals, such as hydrochloric acid, which can contaminate local water sources and harm aquatic ecosystems.

SOCIAL CONCERNS

1.Human Rights: The lithium mining industry in China has been criticized for human rights violations, including the forced relocation of local communities and the use of forced labor.

2.Health Impacts: The release of pollutants from lithium mining and processing can have negative health impacts on nearby communities, including respiratory problems and cancer.

3.Economic Inequality: The benefits of the lithium mining industry are not evenly distributed, and local communities may not receive a fair share of the economic benefits.

4.Cultural Impacts: Lithium mining can also have cultural impacts, as it can disrupt traditional livelihoods and ways of life.

SOLUTIONS AND MITIGATION STRATEGIES

To address these environmental and social concerns, China's government and the lithium mining industry need to take a proactive approach to mitigate the impacts of lithium mining. Some of the strategies that can be employed include:

1.Sustainable Mining Practices: Lithium mining companies can adopt sustainable mining practices, such as using less water and minimizing land disturbance.

2.Environmental Protection Measures: Environmental protection measures, such as the use of protective barriers, can help prevent contamination of water sources and other environmental impacts.

3.Community Engagement: Lithium mining companies can engage with local communities to ensure they are aware of the potential impacts of mining and to develop mutually beneficial solutions.

4.Human Rights Protections: The Chinese government can enforce labor laws and human rights protections to ensure that workers are treated fairly and that communities are not forcibly relocated.

CONCLUSION

China's lithium mining industry provides significant economic benefits, but it also has environmental and social impacts that need to be addressed. By employing sustainable mining practices, implementing environmental protection measures, engaging with local communities, and enforcing human rights protections, the industry can minimize its negative impacts and contribute to the global transition towards clean energy in a responsible and sustainable way.

ENVIRONMENTAL AND SOCIAL CONCERNS SURROUNDING LITHIUM MINING IN CHINA

As the world's largest producer of lithium, China's lithium mining industry has significant environmental and social impacts. Here are some of the main concerns surrounding lithium mining in China:

ENVIRONMENTAL CONCERNS

1.Water Usage: Lithium mining requires significant amounts of water, which can put pressure on already scarce water resources in some regions. This can lead to water shortages and negatively impact local communities and ecosystems.

2.Land Use: Lithium mining often involves large-scale excavation and land disturbance, which can lead to soil erosion, habitat loss, and other environmental impacts.

3.Air Pollution: The processing of lithium ores can release pollutants into the air, such as dust, sulfur dioxide, and carbon dioxide, which can contribute to air pollution and climate change.

4.Chemical Contamination: The processing of lithium ores involves the use of chemicals, such as hydrochloric acid, which can contaminate local water sources and harm aquatic ecosystems.

SOCIAL CONCERNS

1.Human Rights: The lithium mining industry in China has been criticized for human rights violations, including the forced relocation of local communities and the use of forced labor.

2.Health Impacts: The release of pollutants from lithium mining and processing can have negative health impacts on nearby communities, including respiratory problems and cancer.

3.Economic Inequality: The benefits of the lithium mining industry are not evenly distributed, and local communities may not receive a fair share of the economic benefits.

4.Cultural Impacts: Lithium mining can also have cultural impacts, as it can disrupt traditional livelihoods and ways of life.

SOLUTIONS AND MITIGATION STRATEGIES

To address these environmental and social concerns, China's government and the lithium mining industry need to take a proactive approach to mitigate the impacts of lithium mining. Some of the strategies that can be employed include:

1.Sustainable Mining Practices: Lithium mining companies can adopt sustainable mining practices, such as using less water and minimizing land disturbance.

2.Environmental Protection Measures: Environmental protection measures, such as the use of protective barriers, can help prevent contamination of water sources and other environmental impacts.

3.Community Engagement: Lithium mining companies can engage with local communities to ensure they are aware of the potential impacts of mining and to develop mutually beneficial solutions.

4.Human Rights Protections: The Chinese government can enforce labor laws and human rights protections to ensure that workers are treated fairly and that communities are not forcibly relocated.

CONCLUSION
China's lithium mining industry provides significant economic benefits, but it also has

environmental and social impacts that need to be addressed. By employing sustainable mining practices, implementing environmental protection measures, engaging with local communities, and enforcing human rights protections, the industry can minimize its negative impacts and contribute to the global transition towards clean energy in a responsible and sustainable way.

FUTURE OUTLOOK FOR THE CHINESE LITHIUM MINING INDUSTRY

The Chinese lithium mining industry is expected to continue to play a major role in the global lithium market, as the demand for lithium-ion batteries for electric vehicles and energy storage systems continues to grow. Here are some of the key factors that will shape the future outlook for the Chinese lithium mining industry:

1.Government Support: The Chinese government has made a concerted effort to support the development of its lithium mining industry, including providing subsidies and tax incentives for lithium producers and investing in research and development of lithium-related technologies.

2.Technological Innovation: China's lithium mining industry is expected to continue to innovate and improve its technology and processes, leading to increased efficiency and lower costs.

3.Expansion of Production Capacity: Many lithium mining companies in China are planning to expand their production capacity in response to increasing demand for lithium. This includes both domestic and international companies operating in China.

4.Diversification of Supply: China is also exploring other sources of lithium, including from countries such as Australia and Argentina, to diversify its supply and reduce its reliance on domestic production.

5.Environmental and Social Concerns: As the environmental and social impacts of lithium mining become more widely known, there is growing pressure on the Chinese government and lithium mining companies to address these concerns and adopt sustainable practices.

6.Global Competition: While China is currently the largest producer of lithium, other countries such as Australia and Chile are also expanding their lithium production. This may lead to increased competition in the global lithium market.

Overall, the future outlook for the Chinese lithium mining industry looks positive, with significant potential for growth and innovation. However, the industry will need to address the environmental and social impacts of lithium mining and adopt sustainable practices to ensure its long-term viability and success.

THE GLOBAL ADVANTAGES OF THE RELATIONSHIP BETWEEN AUSTRALIA

AND CHINA IN REGARD THE TO MINING
INDUSTRY

The relationship between Australia and China in
the mining industry has many global advantages,
including:

1.Resource Exchange: Australia is rich in natural
resources, including minerals such as iron ore,
coal, and gold. China has a large demand for
these resources to support its manufacturing
industry. The partnership between the two
countries allows for a mutually beneficial
exchange of resources, where China can access
the resources it needs, and Australia can benefit
from the demand for its resources.

2.Economic Benefits: The partnership between
Australia and China in the mining industry has
significant economic benefits for both countries.
For Australia, the mining industry is a major
contributor to its economy, and China is a key
market for its minerals. For China, access to

Australia's resources helps to support its manufacturing industry and economic growth.

3.Trade Relationship: The mining industry is just one part of the broader trade relationship between Australia and China. The two countries have a strong trading partnership, with China being Australia's largest trading partner. The mining industry is a significant part of this relationship, with China being a major market for Australian minerals.

4.Technology Exchange: The mining industry requires advanced technology to extract and process minerals. The partnership between Australia and China in the mining industry allows for an exchange of technology and expertise, benefiting both countries.

5.Environmental Cooperation: The mining industry can have significant environmental impacts, and cooperation between countries can help to mitigate these impacts. Australia and China can work together to develop sustainable mining practices and share knowledge and expertise on environmental management.

Overall, the relationship between Australia and China in the mining industry has many global advantages, including resource exchange, economic benefits, technology exchange, and environmental cooperation. The partnership between the two countries in the mining industry is an important part of their broader trade relationship, and benefits both countries and the global community.

CHAPTER THREE

COMPARISON OF LITHIUM MINING IN AUSTRALIA AND CHINA

Lithium mining is an important industry in both Australia and China, with both countries having significant lithium reserves and contributing to the global supply of the metal. Here is a comparison of the lithium mining industries in Australia and China:

1.Reserves: Australia is currently the world's largest producer of lithium, with an estimated 6.3 million metric tons of reserves. China has the second-largest reserves, with an estimated 3.2 million metric tons.

2.Production: Australia currently produces around 46% of the world's lithium, while China produces around 21%. However, China is rapidly expanding its lithium production capacity and is expected to increase its share of global production.

3.Investment: Both Australia and China are investing in their lithium mining industries. Australia has attracted significant foreign investment in its lithium industry, with companies such as Albemarle and Tianqi investing in Australian lithium mines. China is also investing heavily in its domestic lithium industry, with companies such as Ganfeng Lithium and Tianqi expanding their production capacity.

4.Environmental and Social Concerns: Both Australia and China face environmental and social concerns surrounding their lithium mining industries. In Australia, concerns include water usage and contamination, while in China, concerns include air pollution and labor rights.

5.Technology and Innovation: Both Australia and China are investing in research and development of lithium-related technologies, including improving the efficiency of lithium extraction and battery production.

6.Trade Relationship: China is a major market for Australian lithium, with the majority of Australia's lithium exports going to China. The trade relationship between the two countries is an important aspect of the lithium mining industry.

Overall, the lithium mining industries in Australia and China share many similarities and are both important players in the global lithium market. While Australia currently has a larger share of global production, China is rapidly expanding its production capacity and investing heavily in its domestic lithium industry. The environmental and social concerns surrounding lithium mining are also present in both countries, highlighting the need for sustainable practices in the industry.

PRODUCTION VOLUMES AND TRENDS IN AUSTRALIA AND CHINA

Production volumes and trends in the lithium mining industries in Australia and China are important indicators of the global supply of lithium. Here is a detailed overview of production volumes and trends in both countries:

AUSTRALIA:
Australia is currently the largest producer of lithium, with an estimated 46% of global production in 2021.
The majority of Australian lithium production comes from two mines, Greenbushes in Western Australia and Mount Cattlin in the same region.
Production in Australia has increased significantly in recent years, with an average annual growth rate of 22% between 2015 and 2020.
In 2020, Australia produced approximately 80,000 metric tons of lithium, up from approximately 33,000 metric tons in 2015.
Australia's lithium production is expected to continue to grow, with the Australian government predicting production to reach 220,000 metric tons by 2025.

CHINA:

China is the second-largest producer of lithium, with an estimated 21% of global production in 2021.

The majority of Chinese lithium production comes from brine operations in Tibet, Qinghai, and Sichuan provinces.

Production in China has also increased significantly in recent years, with an average annual growth rate of 23% between 2015 and 2020.

In 2020, China produced approximately 50,000 metric tons of lithium, up from approximately 17,000 metric tons in 2015.

China's lithium production is expected to continue to grow, with some estimates predicting production to reach 80,000 metric tons by 2025.

Production trends:

Both Australia and China are investing heavily in their lithium mining industries and expanding their production capacity.

Australia is expanding its existing mines and developing new mines, while China is increasing its capacity through the development of new brine operations and the expansion of existing ones.

The global demand for lithium is expected to continue to grow, driven by the increasing demand for electric vehicles and energy storage systems.
As a result, both Australia and China are well-positioned to continue to grow their lithium production and meet the growing global demand for the metal.
In conclusion, both Australia and China are significant players in the global lithium mining industry, with production volumes increasing rapidly in recent years. Both countries are investing in expanding their production capacity and are well-positioned to meet the growing global demand for lithium.

THE COST STRUCTURES AND COMPETITIVENESS OF LITHIUM MINING IN AUSTRALIA AND CHINA

The cost structures and competitiveness of lithium mining in Australia and China are important factors that affect their respective positions in the global lithium market. Here is an overview of the cost structures and competitiveness of lithium mining in both countries:

COST STRUCTURES:

The cost structures of lithium mining in Australia and China are different due to the different mining methods used.

In Australia, lithium is mainly mined from hard rock deposits, which require more complex and expensive extraction processes compared to the brine mining operations used in China.

The higher cost of mining in Australia is partially offset by the higher grade of lithium ore found in Australian mines, which typically contain higher levels of lithium per unit of ore.

In China, the brine mining operations have lower production costs, but are subject to greater environmental regulations and restrictions, which can increase operational costs.

COMPETITIVENESS:

Both Australia and China are major players in the global lithium market, but they have different competitive advantages.
Australia has a significant advantage in the production of high-grade lithium ore, which is favored by some battery manufacturers.
China, on the other hand, has a competitive advantage in the production of low-cost lithium products due to the cheaper production costs of brine mining operations.
China also has a large domestic market for lithium products due to its growing electric vehicle and battery manufacturing industries, which gives it a competitive advantage in terms of proximity to its customers.

Both countries are investing heavily in research and development to improve their respective lithium mining processes and technologies to maintain their competitive positions in the global market.

In conclusion, the cost structures and competitiveness of lithium mining in Australia and China are influenced by their respective mining methods and advantages. While Australia has an advantage in high-grade ore production, China has a competitive edge in low-cost brine mining operations and access to a growing domestic market. Both countries are investing in R&D to maintain their competitive positions in the global lithium market as demand for the metal continues to grow.

CHAPTER FOUR

THE ENVIRONMENTAL AND SOCIAL IMPACTS OF LITHIUM MINING IN AUSTRALIA AND CHINA

Lithium mining is an important industry for the production of batteries used in various electronic devices, such as smartphones and electric vehicles. However, it also has significant environmental and social impacts in both Australia and China. Here is an overview of the environmental and social impacts of lithium mining in both countries:

ENVIRONMENTAL IMPACTS:

IN AUSTRALIA:

Lithium mining in Australia is primarily from hard rock deposits, which requires extensive land clearing and can result in habitat loss for native plants and animals.

The mining process can also result in soil erosion, dust and noise pollution, and water contamination if not managed properly.

The use of chemicals in the mining process, such as sulfuric acid and hydrochloric acid, can pose a risk to local ecosystems if they are not properly contained and disposed of.

The mining and production of lithium also requires significant energy inputs, which can lead to increased greenhouse gas emissions and contribute to climate change.

IN CHINA:

Brine mining for lithium is the primary method used in China, which can have significant impacts on water resources.

Large volumes of water are pumped from underground reservoirs to extract lithium, which can lead to a reduction in the water table and affect the availability of water for local communities and agriculture.

The use of chemicals in the brine mining process can also lead to water contamination if not properly managed and disposed of.

In addition, the mining and processing of lithium in China requires significant energy inputs, which can contribute to air pollution and greenhouse gas emissions.

Social impacts:

IN AUSTRALIA:

Lithium mining can have significant impacts on local communities, particularly Indigenous communities, who may have cultural and spiritual connections to the land being mined.

The mining industry can also create social disruptions, such as increased traffic, noise, and dust, which can affect the quality of life for nearby residents.

The influx of workers to mining communities can also place pressure on local services, such as housing and healthcare.

IN CHINA:

The brine mining industry in China has been associated with social conflicts between mining companies and local communities over land rights and compensation for land use.

The mining industry can also create social disruptions, such as increased traffic, noise, and dust, which can affect the quality of life for nearby residents.

The working conditions in some lithium mines in China have also been criticized for their lack of safety standards and labor rights.

In conclusion, lithium mining in both Australia and China has significant environmental and social impacts. These impacts include habitat loss, water contamination, air pollution, and social disruptions. It is important for mining companies and governments to prioritize sustainable practices and engage with local communities to mitigate these impacts and ensure the long-term sustainability of the industry.

THE REGULATORY FRAMEWORKS FOR LITHIUM MINING IN AUSTRALIA AND CHINA

Lithium mining is a highly regulated industry in both Australia and China. Here is an overview of the regulatory frameworks for lithium mining in each country:

AUSTRALIA:

The Australian government has a federal regulatory framework for mining activities, which is overseen by the Department of Industry, Science, Energy and Resources.
State and territory governments also have their own regulatory frameworks for mining activities, which cover issues such as environmental management, health and safety, and land access.
The Australian government has also established a National Lithium Strategy to support the development of the lithium industry in Australia, which includes initiatives to improve regulatory frameworks and support research and development.

CHINA:
The Chinese government has a complex regulatory framework for mining activities,

which includes multiple agencies responsible for different aspects of regulation.

The Ministry of Natural Resources is responsible for issuing mining licenses and regulating mining activities, while the Ministry of Ecology and Environment is responsible for environmental protection.

In addition, local governments have their own regulatory frameworks for mining activities, which can vary depending on the region.

The Chinese government has also established a range of policies and regulations to support the development of the lithium industry, including subsidies for lithium production and research and development programs.

Overall, both Australia and China have regulatory frameworks in place to manage the environmental and social impacts of lithium mining activities. However, there have been concerns raised about the effectiveness of these frameworks in both countries, particularly around issues such as water management and community engagement. It is important for governments and mining companies to work together to ensure that regulatory frameworks are effective and responsive to the needs of local communities and the environment.

THE EFFECTS OF GOVERNMENT POLICIES ON LITHIUM MINING IN BOTH COUNTRIES

Government policies play a significant role in shaping the lithium mining industry in both Australia and China. Here is an overview of the effects of government policies on lithium mining in each country:

AUSTRALIA:

The Australian government has implemented policies to support the development of the lithium mining industry, including the National Lithium Strategy.
This strategy includes initiatives to improve the regulatory framework for the industry, support

research and development, and promote investment in lithium projects.

The Australian government has also established a range of subsidies and tax incentives to support the development of the lithium industry.

CHINA:

The Chinese government has implemented a range of policies to support the development of the lithium mining industry, including subsidies for lithium production and research and development programs.

The government has also established regulations to control the amount of lithium produced and exported from the country, in order to maintain a stable lithium market and protect national resources.

In addition, the Chinese government has implemented policies to encourage the development of electric vehicles, which has increased demand for lithium and other battery minerals.

Overall, government policies have had a significant impact on the development of the lithium mining industry in both Australia and China. These policies have supported the growth of the industry, but they have also been criticized for their potential environmental and social impacts. It is important for governments to balance economic development with environmental and social sustainability, and to ensure that the benefits of the industry are shared equitably among all stakeholders.

HIGHLIGHTS OF THE POLICIES

Here are some of the key policies that have had an impact on the lithium mining industry in Australia and China:

AUSTRALIA:

1.National Lithium Strategy: In 2019, the Australian government launched the National Lithium Strategy, which includes a range of initiatives to support the development of the lithium industry. These initiatives include improving the regulatory framework, promoting investment in lithium projects, and supporting research and development.

2.Investment incentives: The Australian government has established a range of subsidies and tax incentives to support the development of the lithium industry. These include the Exploration Development Incentive, which provides tax incentives for exploration activities, and the Junior Minerals Exploration Incentive, which provides funding for exploration activities.

3.Environmental regulations: The Australian government has established environmental regulations to manage the impacts of lithium mining activities. These regulations cover issues such as land access, water management, and rehabilitation requirements.

CHINA:

Subsidies for lithium production: The Chinese government has implemented a range of subsidies to support the production of lithium in the country. These subsidies are designed to encourage the development of the lithium industry and increase domestic production of lithium.

1.Research and development programs: The Chinese government has established research and development programs to support the development of the lithium industry. These programs focus on improving production processes, developing new technologies, and promoting the use of lithium in high-tech industries.

2.Export controls: The Chinese government has implemented regulations to control the amount of lithium produced and exported from the country. These regulations are designed to maintain a stable lithium market and protect national resources.

3.Electric vehicle subsidies: The Chinese government has implemented policies to encourage the development of electric vehicles, which has increased demand for lithium and other battery minerals. These policies include subsidies for electric vehicle manufacturers and incentives for consumers to purchase electric vehicles.

Overall, these policies have had a significant impact on the development of the lithium mining industry in both Australia and China. While they have provided important support for the industry, there are also concerns about their potential environmental and social impacts. It is important for governments to continue to evaluate and refine their policies to ensure that they balance economic development with environmental and social sustainability.

CHAPTER FIVE

TECHNOLOGICAL DEVELOPMENTS AND INNOVATIONS IN LITHIUM MINING IN AUSTRALIA AND CHINA

Lithium mining is a rapidly evolving industry, and both Australia and China have been at the forefront of technological developments and innovations in the field. Here are some of the key developments in lithium mining technology in these countries:

AUSTRALIA:

1.Direct lithium extraction (DLE): DLE is a technology that extracts lithium from brines using chemical processes. This technology has the potential to significantly reduce the cost and environmental impact of lithium extraction compared to traditional evaporation methods. Australian companies, such as Lithium Australia and Lake Resources, have been at the forefront of developing and commercializing DLE technology.

2.Lithium processing: Australia has also been investing in the development of processing technologies to produce high-purity lithium products for use in batteries. This includes the development of techniques to remove impurities from lithium concentrates, and the production of lithium hydroxide, which is a key input in battery production.

CHINA:

Lithium-ion battery technology: China has invested heavily in the development of lithium-ion battery technology, which is a key application for lithium. Chinese companies such as CATL and BYD are among the world's largest producers of lithium-ion batteries, and China accounts for a significant share of global battery production.

1.Recycling technologies: China has also been developing recycling technologies for lithium-ion batteries, which can help to recover valuable metals such as lithium, cobalt, and nickel. This can reduce the environmental impact of mining new materials and ensure the security of supply for these critical minerals.

2.New mining methods: Chinese companies have been exploring new methods for lithium mining, including in-situ leaching (ISL) and solution mining. These methods involve injecting solutions into underground formations to extract lithium, which can reduce the need for open-pit mining and other traditional mining techniques.

Overall, technological developments and innovations in lithium mining in both Australia and China have the potential to significantly improve the efficiency and sustainability of the industry. Continued investment in research and development will be key to driving further progress in the field, while also addressing the environmental and social concerns associated with lithium mining.

THE FUTURE OF LITHIUM MINING IN AUSTRALIA AND CHINA

Lithium mining is a rapidly growing industry, driven by the increasing demand for lithium-ion batteries used in electric vehicles, consumer electronics, and energy storage systems. Australia and China are two of the world's largest producers of lithium, and the future of lithium mining in these countries looks bright. Here are some of the key trends and developments shaping the future of the industry in Australia and China.

AUSTRALIA:

1.Increasing production: Australia is already the world's largest producer of lithium, with several major mines in operation. However, production is expected to increase significantly in the coming years, with new mines and expansions planned. For example, the Greenbushes mine in Western Australia, which is one of the world's largest lithium mines, is undergoing a major expansion that will double its production capacity.

2 Exploration and development of new resources: In addition to expanding existing mines, Australian companies are also exploring new lithium resources. This includes both traditional hard rock deposits and unconventional sources such as brines and clays. Companies such as Kidman Resources and Pilbara Minerals are investing in the development of new lithium resources in Australia.

3.Investment in downstream processing: Australia is also investing in the development of downstream processing facilities to produce high-purity lithium products for use in batteries. For example, the Western Australian government has announced plans to build a $1.3 billion lithium processing plant that will produce lithium hydroxide and other battery materials.

CHINA:

1.Growing demand for lithium-ion batteries: China is the world's largest market for electric vehicles, and this is driving significant demand for lithium-ion batteries. Chinese companies are investing heavily in the production of batteries and related technologies, and the government has announced plans to phase out the production and sale of traditional fossil-fueled vehicles in the coming years.

2.Expansion of mining operations: China is also expanding its lithium mining operations, with several new mines in development. For example, the Qinghai Salt Lake Industry Group, which is one of China's largest lithium producers, is building a new lithium mine in Tibet that is expected to begin production in 2022.

3.Investment in recycling technologies: China is also investing in the development of recycling technologies for lithium-ion batteries. This can help to reduce the environmental impact of mining new materials and ensure the security of supply for critical minerals. For example, Chinese battery manufacturer CATL has announced plans to build a recycling plant that can process up to 120,000 tons of batteries per year.

Overall, the future of lithium mining in Australia and China looks bright, with significant growth expected in production, exploration, and downstream processing. Continued investment in research and development will be key to driving further progress in the field, while also addressing the environmental and social concerns associated with lithium mining. As the world transitions to a low-carbon economy, lithium will play an increasingly important role in powering the technologies of the future, and Australia and China will be at the forefront of this industry.

CHINA'S NEW LITHIUM MINE IN TIBET

China's new lithium mine in Tibet is a significant development in the country's push to secure its supply of key minerals and materials for its rapidly growing electric vehicle industry. The mine, which is being developed by the Qinghai Salt Lake Industry Group, is located in the Lhunze County of Tibet and is expected to begin production in 2022.

The mine is estimated to have reserves of around 500,000 tons of lithium, making it one of the largest lithium deposits in China. The development of the mine is part of China's broader efforts to ramp up its domestic production of lithium, which is a critical component of the batteries used in electric vehicles.

The project has been controversial due to its location in a sensitive border region, and concerns have been raised about potential environmental and social impacts. Lhunze County is located in a high-altitude area with fragile ecosystems, and there are concerns that the mining operations could damage the local environment and threaten the livelihoods of local communities.

The Chinese government has acknowledged these concerns and has promised to implement measures to mitigate the environmental and social impacts of the project. The Qinghai Salt Lake Industry Group has also committed to using advanced mining technologies and best practices to minimize the environmental impact of the project.

Despite the concerns, the development of the lithium mine in Tibet is a significant step forward for China's efforts to secure its supply of key minerals and materials. With the country's electric vehicle industry continuing to grow rapidly, the demand for lithium is expected to increase significantly in the coming years. By developing new mines and expanding existing operations, China is positioning itself to become a major player in the global lithium market.

CHAPTER SIX

THE EMERGING TRENDS AND CHALLENGES IN THE LITHIUM MINING INDUSTRY

The lithium mining industry has experienced significant growth over the past decade due to the increasing demand for lithium-ion batteries used in electric vehicles (EVs), renewable energy storage, and consumer electronics. However, as demand continues to rise, the industry is also facing emerging trends and challenges that will shape its future.

EMERGING TRENDS:

1.Increased demand for lithium: The demand for lithium is expected to grow exponentially due to the increasing adoption of EVs and the shift towards renewable energy. According to some estimates, the global demand for lithium could triple by 2025.

2.Growth in lithium recycling: As the demand for lithium grows, there is a growing need to recycle the metal from used batteries to meet the demand sustainably. Recycling is a cost-effective way to reduce the environmental impact of mining and ensure a steady supply of lithium.

3.New sources of lithium: The traditional sources of lithium, such as brine and hard rock mining, are being supplemented by new sources such as geothermal brines and oilfield brines. This is opening up new avenues for lithium production.

4.The rise of vertically integrated mining companies: Some lithium mining companies are becoming vertically integrated, meaning they are controlling the entire supply chain from mining to battery production. This helps them to optimize costs and ensure a stable supply of lithium.

CHALLENGES:

1.Environmental impact: Lithium mining has significant environmental impacts, such as water depletion and pollution, land degradation, and greenhouse gas emissions. Addressing these issues requires careful planning and management to minimize the impact on the environment and local communities.

2.Supply chain issues: The lithium supply chain is complex and involves multiple stakeholders, including miners, refiners, and battery manufacturers. Ensuring a reliable and stable supply chain requires collaboration and coordination between these stakeholders.

3.High capital costs: Lithium mining requires significant upfront capital investments, which can make it challenging for smaller companies to enter the market. This also makes it challenging to scale up production quickly to meet growing demand.

4.Geopolitical risks: Most of the world's lithium reserves are concentrated in a few countries, including Chile, Argentina, and Australia. This creates geopolitical risks that could disrupt the supply chain and affect prices.

In conclusion, the lithium mining industry is facing both opportunities and challenges as demand for lithium grows. Addressing the environmental and social impacts of mining, improving the supply chain, and managing capital costs will be key to ensuring sustainable and profitable growth in the industry.

THE OPPORTUNITIES FOR GROWTH AND INVESTMENT IN THE AUSTRALIAN AND CHINESE LITHIUM MINING INDUSTRIES

The Australian and Chinese lithium mining industries offer significant opportunities for growth and investment due to their rich mineral resources and increasing demand for lithium. Here are some of the key factors that make these two countries attractive for investors:

OPPORTUNITIES IN THE AUSTRALIAN LITHIUM MINING INDUSTRY:

Abundant mineral resources: Australia has some of the largest lithium reserves in the world, with deposits located in Western Australia and Queensland. This gives the country a competitive advantage in the global lithium market.

1.Favorable investment environment: Australia has a stable political and economic environment, with well-established legal systems and transparent regulatory frameworks. This makes it an attractive destination for foreign investors.

2.Growing demand for lithium: The demand for lithium is expected to grow exponentially due to the increasing adoption of electric vehicles and renewable energy storage. Australia is well-positioned to capitalize on this trend due to its abundant reserves and proximity to Asian markets.

3.Strong government support: The Australian government has been supportive of the lithium mining industry, providing funding for exploration and research to encourage the development of new mines and technologies.

Opportunities in the Chinese Lithium Mining Industry:

4.Largest consumer market: China is the world's largest consumer market for lithium, accounting for more than half of global demand. This makes it a crucial market for lithium producers.

5.Strong government support: The Chinese government has made lithium a priority industry, providing subsidies and other incentives to encourage the development of domestic lithium resources.

6.Abundant mineral resources: China has significant lithium reserves, particularly in the Sichuan, Qinghai, and Tibet regions. This provides a reliable domestic supply of lithium and reduces the country's dependence on imports.

7.Growing demand for electric vehicles: China is the world's largest market for electric vehicles, and this trend is expected to continue. This will drive demand for lithium-ion batteries and create opportunities for lithium producers.

In conclusion, the Australian and Chinese lithium mining industries offer significant opportunities for growth and investment due to their abundant mineral resources, growing demand for lithium, and supportive government policies. Investors should carefully evaluate the risks and opportunities of each market and develop a comprehensive investment strategy that takes into account the unique characteristics of each country.

ALTERNATIVE SOURCES OF LITHIUM AND THEIR POTENTIAL IMPACTS ON THE INDUSTRY

Lithium is a key component in the manufacturing of lithium-ion batteries, which are used in a wide range of devices such as smartphones, electric vehicles, and energy storage systems. The increasing demand for lithium has led to concerns about the availability and sustainability of its primary sources. Currently, the majority of the world's lithium is extracted from brine deposits in South America, primarily in Chile, Argentina, and Bolivia, and hard rock mining in Australia. However, alternative sources of lithium are being explored that could have significant impacts on the industry.

One potential alternative source of lithium is geothermal brines, which are hot water reservoirs found in volcanic regions. These brines contain high concentrations of lithium, as well as other minerals such as potassium and boron. Geothermal brine extraction has the potential to be a more sustainable and environmentally friendly method of lithium extraction compared to traditional mining techniques. In addition, geothermal energy can be generated from the same wells that extract the brine, providing a renewable energy source.

Another potential source of lithium is from seawater. The concentration of lithium in seawater is much lower than in brines, but the vast quantity of seawater available means that there is a significant amount of lithium that can be extracted. However, the process of extracting lithium from seawater is currently very energy-intensive and expensive, making it a less attractive option compared to other sources of lithium.

Another alternative source of lithium is from recycled batteries. Lithium-ion batteries can be recycled, and the recovered lithium can be used to produce new batteries. Recycling lithium from batteries can help reduce the demand for primary sources of lithium, as well as reduce the environmental impact of battery production and disposal.

The development of alternative sources of lithium has the potential to significantly impact the lithium industry. If these sources are successfully developed and scaled up, they could increase the supply of lithium, reduce the cost of lithium production, and reduce the environmental impact of lithium extraction. However, there are also potential challenges associated with developing these sources. For example, geothermal brine extraction could potentially impact local water resources and ecosystems, and the high energy requirements of seawater extraction could make it financially unviable.

In conclusion, alternative sources of lithium such as geothermal brines, seawater, and recycled batteries have the potential to significantly impact the lithium industry. While there are challenges associated with developing these sources, they offer opportunities to increase the sustainability and reduce the environmental impact of lithium production. As demand for lithium continues to grow, it will be important to explore and develop these alternative sources to ensure a stable and sustainable supply of lithium for the future.

THE PROSPECTS FOR SUSTAINABLE AND RESPONSIBLE LITHIUM MINING IN AUSTRALIA AND CHINA

Lithium mining is an essential component of the clean energy transition, with lithium-ion batteries used to power electric vehicles, store renewable energy, and other applications. As the demand for lithium continues to grow, there is a need to ensure that the mining of this critical mineral is done in a sustainable and responsible manner. This article will explore the prospects for sustainable and responsible lithium mining in Australia and China, two of the world's largest lithium producers.

Australia is currently the world's largest lithium producer, with the majority of its production coming from hard rock mining. Australia's lithium mining industry has been developed with a strong focus on sustainability and environmental responsibility, with a well-established regulatory framework in place to ensure that mining activities are conducted in a responsible manner. This includes strict environmental impact assessments, requirements for mine site rehabilitation, and regulations around water and air pollution. Additionally, Australia has strong labor laws and mining industry standards that promote the rights of workers and ensure that they are treated fairly.

In recent years, there has been a growing interest in developing Australia's brine deposits, which are located mainly in Western Australia. Brine deposits have a lower environmental footprint than hard rock mining, as they require less energy and produce less waste. However, there are concerns around water usage, as the extraction of lithium from brines requires large volumes of water, which could potentially impact local ecosystems and water resources.

China is the second-largest producer of lithium, with the majority of its production coming from brine deposits in the Qinghai region. China has been investing heavily in its lithium mining industry, with a focus on developing new technologies to increase efficiency and reduce the environmental impact of mining activities. However, there have been concerns around the environmental and social impacts of lithium mining in China, particularly in relation to water usage and air pollution. In addition, there have been reports of poor working conditions and labor rights violations in some Chinese lithium mines.

To ensure sustainable and responsible lithium mining in Australia and China, there are several steps that can be taken. These include improving water management practices, reducing the environmental impact of mining activities, and promoting labor rights and fair working conditions. Governments and industry stakeholders can also work together to establish clear regulatory frameworks that promote sustainable mining practices and ensure that companies are held accountable for any negative impacts of their activities. Additionally, there is a need to invest in research and development of new technologies to improve the efficiency and sustainability of lithium mining activities.

In conclusion, while there are challenges associated with lithium mining in Australia and China, there are also opportunities to develop a sustainable and responsible lithium mining industry. By working together to improve mining practices, reduce the environmental footprint of mining activities, and promote fair working conditions, the lithium industry can play a crucial role in the transition to a low-carbon economy.

ALTERNATIVE SOURCES OF LITHIUM

Alternative sources of lithium and their potential impacts on the industry

Lithium is a key component in the manufacturing of lithium-ion batteries, which are used in a wide range of devices such as smartphones, electric vehicles, and energy storage systems. The increasing demand for lithium has led to concerns about the availability and sustainability of its primary sources. Currently, the majority of the world's lithium is extracted from brine deposits in South America, primarily in Chile, Argentina, and Bolivia, and hard rock mining in Australia. However, alternative sources of lithium are being explored that could have significant impacts on the industry.

One potential alternative source of lithium is geothermal brines, which are hot water reservoirs found in volcanic regions. These brines contain high concentrations of lithium, as well as other minerals such as potassium and boron. Geothermal brine extraction has the potential to be a more sustainable and environmentally friendly method of lithium extraction compared to traditional mining techniques. In addition, geothermal energy can be generated from the same wells that extract the brine, providing a renewable energy source.

Another potential source of lithium is from seawater. The concentration of lithium in seawater is much lower than in brines, but the vast quantity of seawater available means that there is a significant amount of lithium that can be extracted. However, the process of extracting lithium from seawater is currently very energy-intensive and expensive, making it a less attractive option compared to other sources of lithium.

Another alternative source of lithium is from recycled batteries. Lithium-ion batteries can be recycled, and the recovered lithium can be used to produce new batteries. Recycling lithium from batteries can help reduce the demand for primary sources of lithium, as well as reduce the environmental impact of battery production and disposal.

The development of alternative sources of lithium has the potential to significantly impact the lithium industry. If these sources are successfully developed and scaled up, they could increase the supply of lithium, reduce the cost of lithium production, and reduce the environmental impact of lithium extraction. However, there are also potential challenges associated with developing these sources. For example, geothermal brine extraction could potentially impact local water resources and ecosystems, and the high energy requirements of seawater extraction could make it financially unviable.

In conclusion, alternative sources of lithium such as geothermal brines, seawater, and recycled batteries have the potential to significantly impact the lithium industry. While there are challenges associated with developing these sources, they offer opportunities to increase the sustainability and reduce the environmental impact of lithium production. As demand for lithium continues to grow, it will be important to explore and develop these alternative sources to ensure a stable and sustainable supply of lithium for the future.

PROSPECTS FOR SUSTAINABLE AND RESPONSIBLE LITHIUM MINING IN AUSTRALIA AND CHINA

Lithium mining is an essential component of the clean energy transition, with lithium-ion batteries used to power electric vehicles, store renewable energy, and other applications. As the demand for lithium continues to grow, there is a need to ensure that the mining of this critical mineral is done in a sustainable and responsible manner. This article will explore the prospects for sustainable and responsible lithium mining in Australia and China, two of the world's largest lithium producers.

Australia is currently the world's largest lithium producer, with the majority of its production coming from hard rock mining. Australia's lithium mining industry has been developed with a strong focus on sustainability and environmental responsibility, with a well-established regulatory framework in place to ensure that mining activities are conducted in a responsible manner. This includes strict environmental impact assessments, requirements for mine site rehabilitation, and regulations around water and air pollution. Additionally, Australia has strong labor laws and mining industry standards that promote the rights of workers and ensure that they are treated fairly.

In recent years, there has been a growing interest in developing Australia's brine deposits, which are located mainly in Western Australia. Brine deposits have a lower environmental footprint than hard rock mining, as they require less energy and produce less waste. However, there are concerns around water usage, as the extraction of lithium from brines requires large volumes of water, which could potentially impact local ecosystems and water resources.

China is the second-largest producer of lithium, with the majority of its production coming from brine deposits in the Qinghai region. China has been investing heavily in its lithium mining industry, with a focus on developing new technologies to increase efficiency and reduce the environmental impact of mining activities. However, there have been concerns around the environmental and social impacts of lithium mining in China, particularly in relation to water usage and air pollution. In addition, there have been reports of poor working conditions and labor rights violations in some Chinese lithium mines.

To ensure sustainable and responsible lithium mining in Australia and China, there are several steps that can be taken. These include improving water management practices, reducing the environmental impact of mining activities, and promoting labor rights and fair working conditions. Governments and industry stakeholders can also work together to establish clear regulatory frameworks that promote sustainable mining practices and ensure that companies are held accountable for any negative impacts of their activities. Additionally, there is a need to invest in research and development of new technologies to improve the efficiency and sustainability of lithium mining activities.

In conclusion, while there are challenges associated with lithium mining in Australia and China, there are also opportunities to develop a sustainable and responsible lithium mining industry. By working together to improve mining practices, reduce the environmental footprint of mining activities, and promote fair working conditions, the lithium industry can play a crucial role in the transition to a low-carbon economy.

CONCLUSION

SUMMARY OF KEY FINDINGS

Key findings regarding lithium mining activities in Australia and China include:

AUSTRALIA:

Majority of lithium production comes from hard rock mining, but there is growing interest in developing brine deposits.

Strong regulatory framework in place to ensure responsible and sustainable mining practices.

Concerns around water usage in lithium extraction from brines.

CHINA:

Majority of lithium production comes from brine deposits in the Qinghai region.

Chinese government has been investing heavily in lithium mining industry and developing new technologies.

Concerns around environmental and social impacts of lithium mining, including water usage, air pollution, and labor rights violations.

Overall, while both countries face challenges in their lithium mining activities, there is potential to develop sustainable and responsible mining practices through improved water management, reduced environmental impacts, and promoting fair working conditions.

IMPLICATIONS FOR POLICY MAKERS, INVESTORS, AND INDUSTRY STAKEHOLDERS

The growth of the lithium industry and the increasing demand for lithium-ion batteries have significant implications for policy makers, investors, and industry stakeholders.

Policy makers have an important role to play in promoting sustainable and responsible lithium mining practices. They can establish clear regulatory frameworks that promote sustainable mining practices, including water management, environmental protection, and labor rights. They can also incentivize the development of new technologies to reduce the environmental impact of lithium mining activities. Additionally, policy makers can encourage collaboration between governments, industry stakeholders, and local communities to ensure that the benefits of lithium mining are shared fairly and equitably.

Investors also have a crucial role to play in promoting sustainable lithium mining practices. They can prioritize investments in companies that demonstrate a commitment to sustainability and responsible mining practices. Investors can also engage with mining companies to encourage them to adopt sustainable mining practices and disclose information about their environmental and social impacts. By investing in sustainable mining practices, investors can help to ensure the long-term viability of the lithium industry.

Industry stakeholders, including mining companies, have a responsibility to ensure that their activities are conducted in a sustainable and responsible manner. This includes implementing best practices for water management, environmental protection, and labor rights. Mining companies can also invest in research and development to improve the efficiency and sustainability of their mining activities. Additionally, industry stakeholders can engage with local communities to ensure that the benefits of lithium mining are shared fairly and equitably.

In conclusion, the growth of the lithium industry and the increasing demand for lithium-ion batteries present significant opportunities and challenges for policy makers, investors, and industry stakeholders. By promoting sustainable and responsible mining practices, these stakeholders can help to ensure that the benefits of the lithium industry are realized while minimizing its negative impacts on the environment and local communities.

FUTURE PROSPECTS FOR THE LITHIUM MINING INDUSTRY IN AUSTRALIA AND CHINA

The future prospects for the lithium mining industry in Australia and China are promising, given the increasing demand for lithium-ion batteries and the transition towards a low-carbon economy.

IN AUSTRALIA,
there is a growing interest in developing brine deposits, which have a lower environmental footprint than hard rock mining. However, there are concerns around water usage and potential impacts on local ecosystems and water resources. To address these concerns, there is a need for continued research and development to improve water management practices and reduce the environmental impact of brine extraction.

In addition, Australia has a well-established regulatory framework in place to ensure that mining activities are conducted in a responsible and sustainable manner. This regulatory framework provides a strong foundation for the development of a sustainable and responsible lithium mining industry, which could play a significant role in the country's transition towards a low-carbon economy.

IN CHINA,

the government has been investing heavily in the lithium mining industry and developing new technologies to increase efficiency and reduce the environmental impact of mining activities. However, there have been concerns around environmental and social impacts of lithium mining, particularly in relation to water usage and air pollution. To address these concerns, there is a need for continued investment in research and development of new technologies to improve the sustainability of mining activities.

Despite these challenges, the growing demand for lithium-ion batteries presents significant opportunities for the lithium mining industry in both Australia and China. With the right investments in sustainable and responsible mining practices, the industry could play a crucial role in the transition to a low-carbon economy, providing a secure and reliable source of lithium for use in electric vehicles, renewable energy storage, and other applications.

Overall, the future prospects for the lithium mining industry in Australia and China are promising, provided that there is continued investment in sustainable and responsible mining practices, regulatory frameworks that promote environmental protection and labor rights, and research and development of new technologies to improve the efficiency and sustainability of mining activities.